A Bug Band

by Lynnette Timothy
illustrated by Ken Bowser

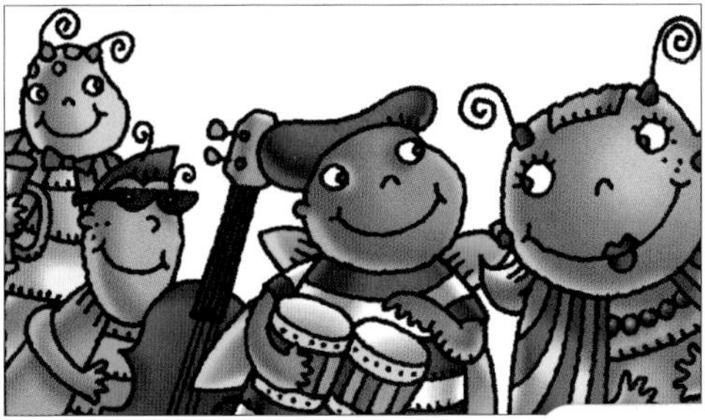

D1708670

Harcourt

Orlando Boston Dallas Chicago San Diego

www.harcourtschool.com

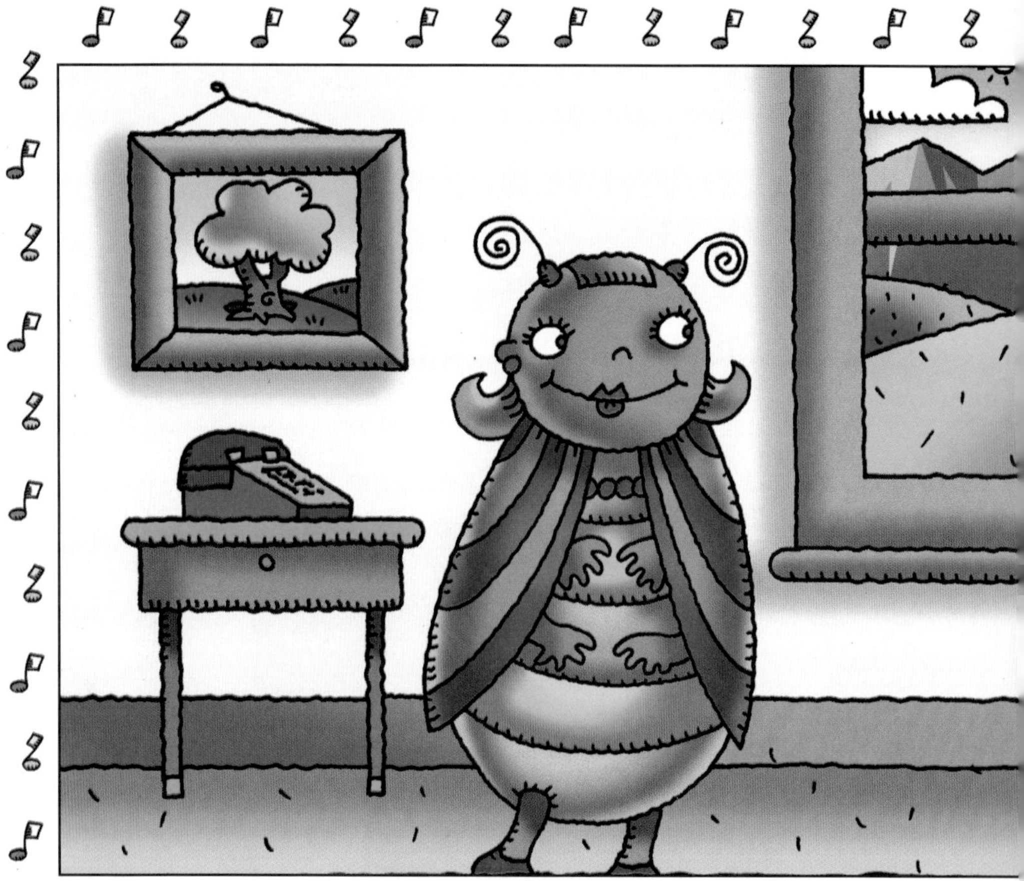

I bug

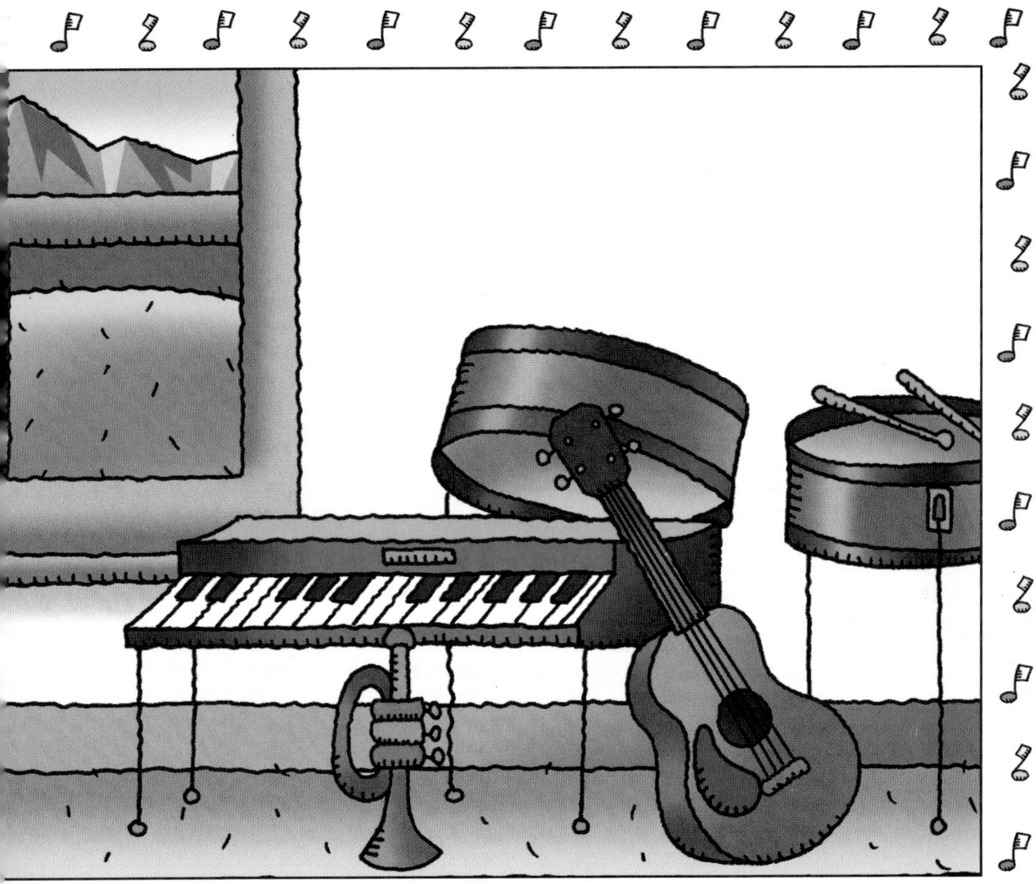

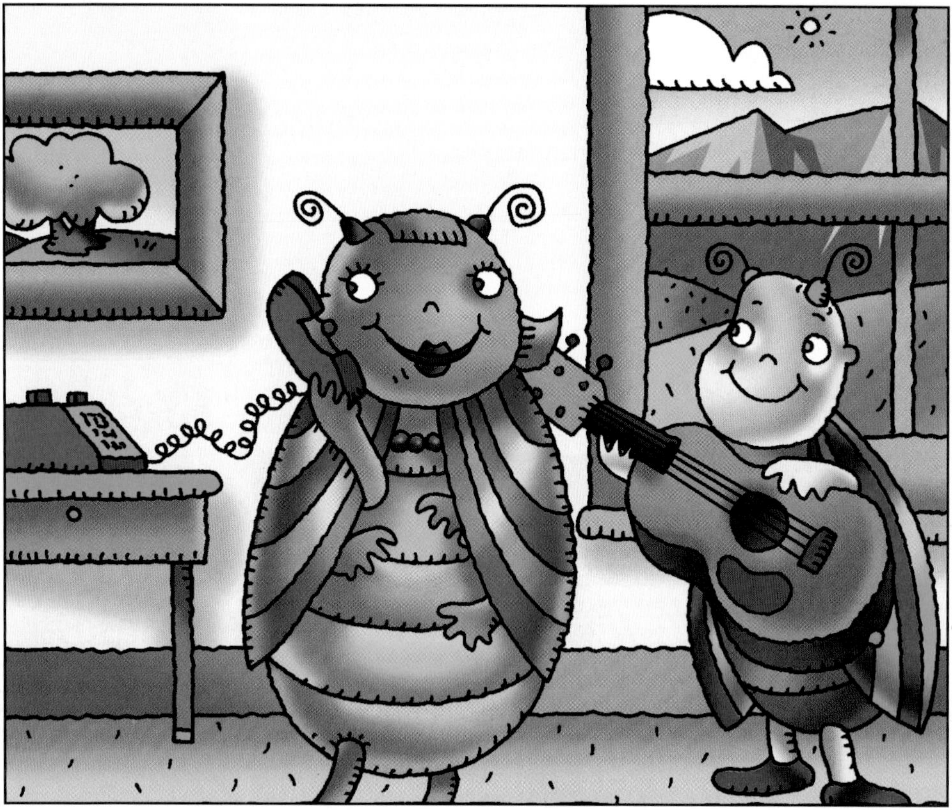

2 bugs

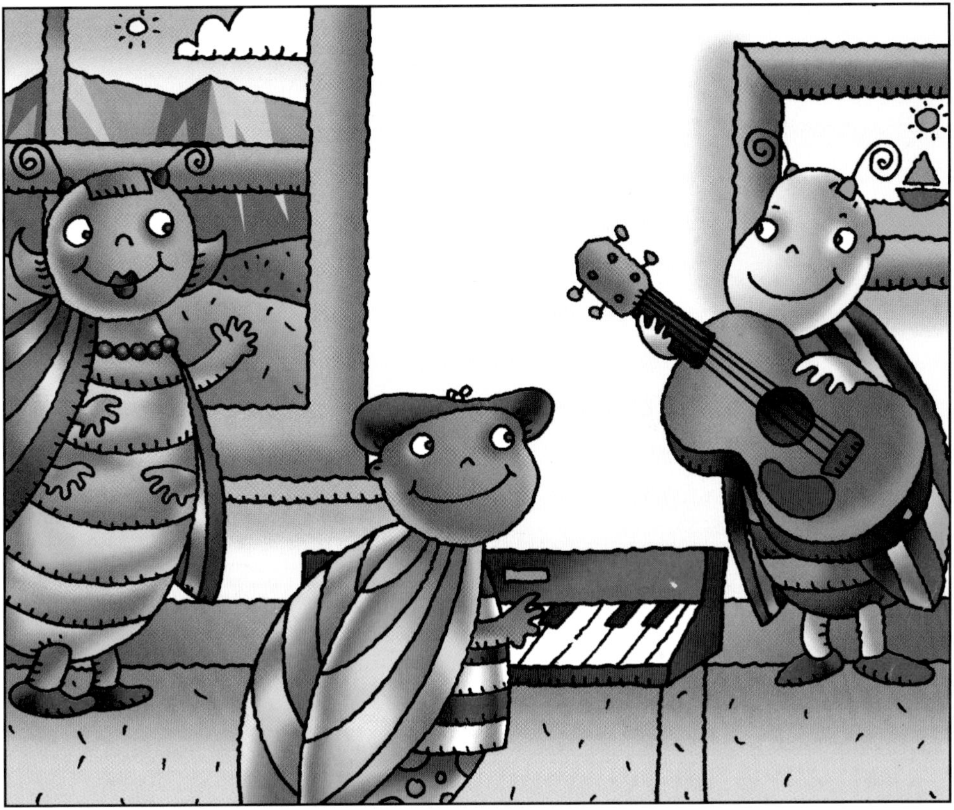

3 bugs

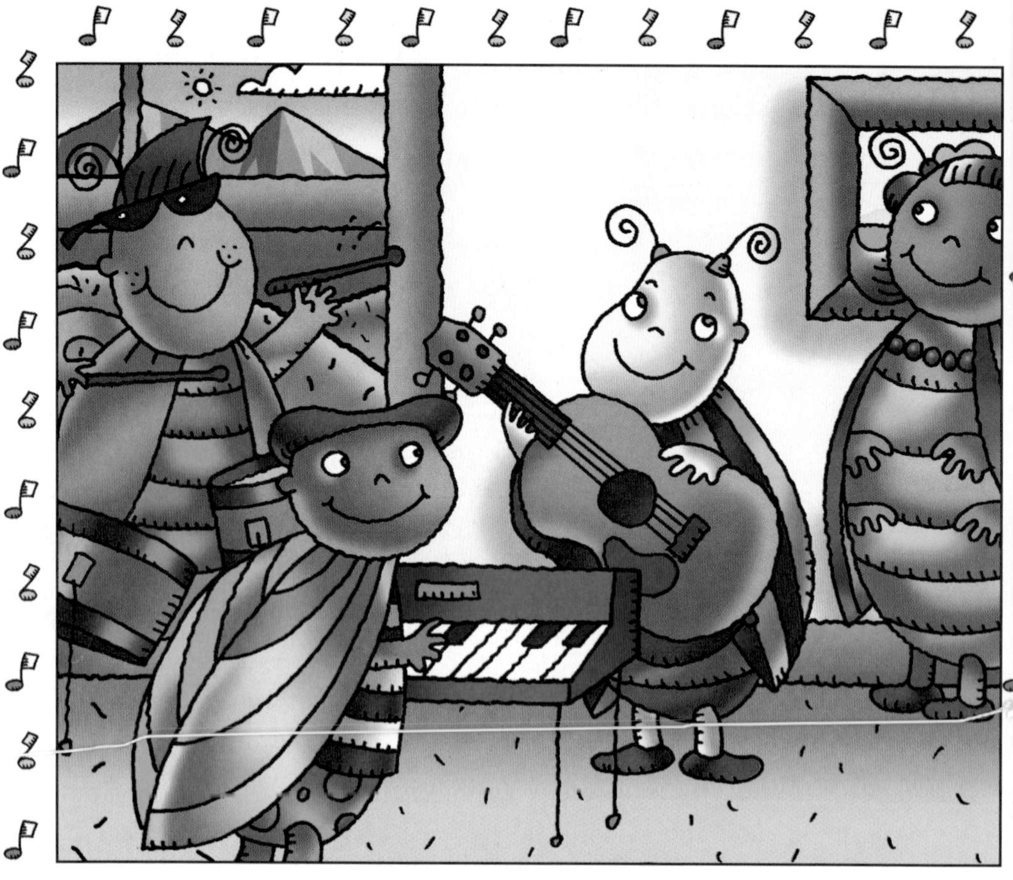

4 bugs

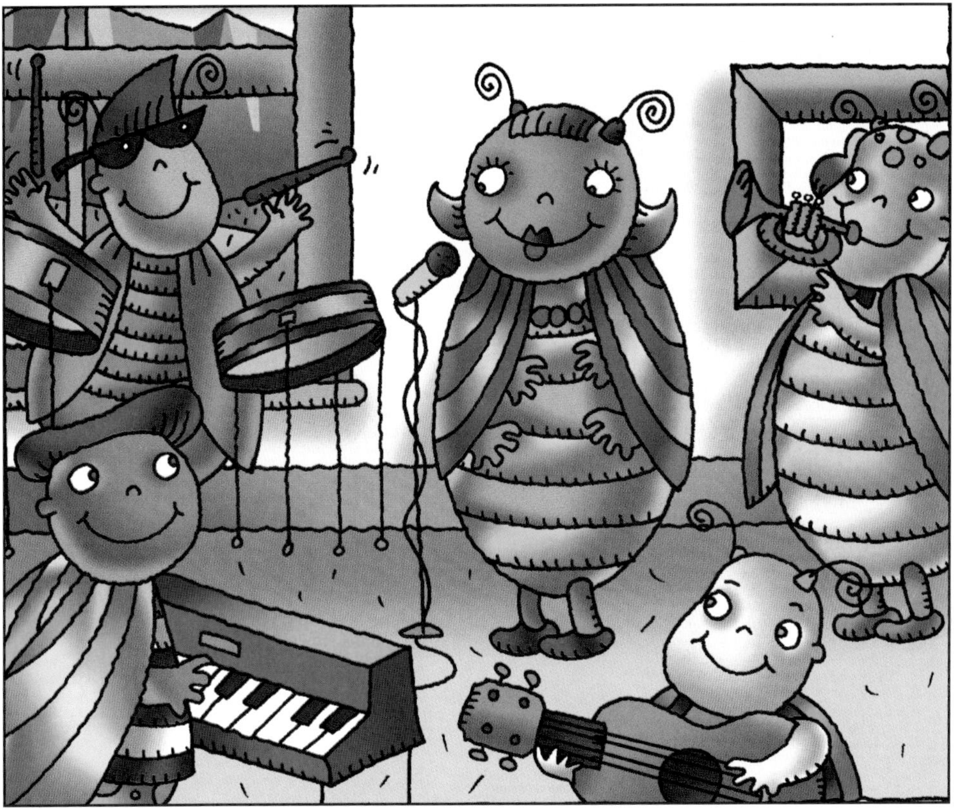

5 bugs

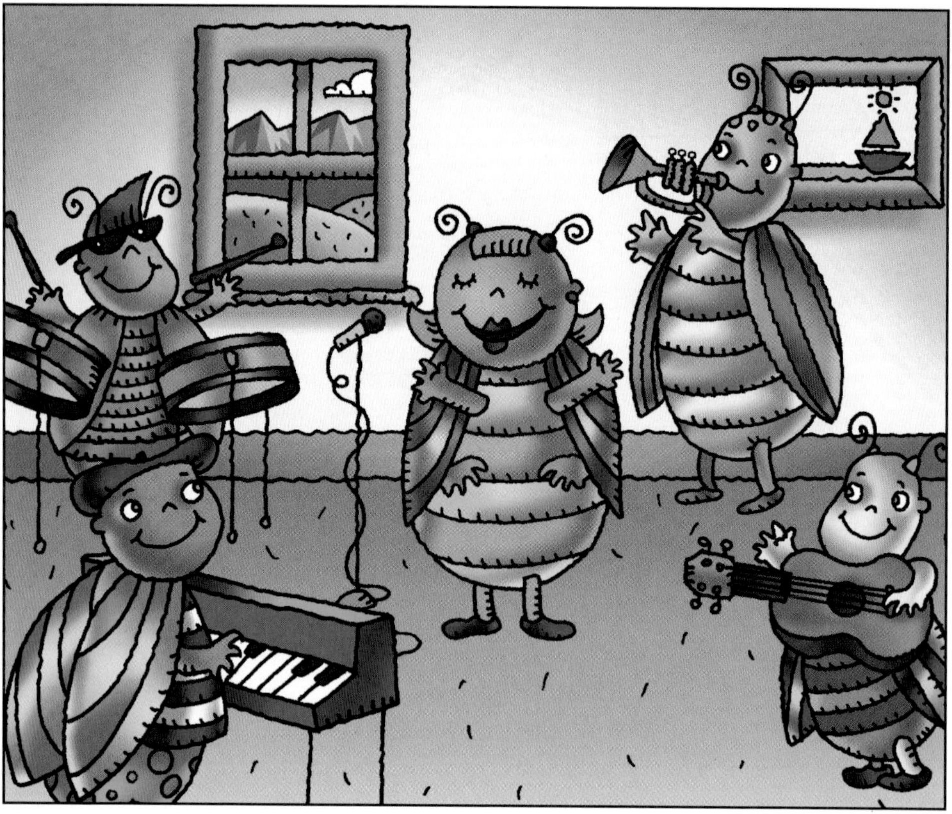

A bug band!